WILD WASHINGTON

AMAZING WILDLIFE IN AND AROUND OUR NATION'S CAPITAL

Photographs and Text by

Don Chernoff

www.dcwild.com

To Lucy, Josie, and Max

Wild Washington™ 2004

Published by Don Chernoff
www.dcwild.com
don@dcwild.com

First Printing 2005

Printed in China

Library of Congress Catalog Card Number 2005906244

ISBN 1-933570-14-8

Design and Layout by Vivo Design

There are many people who contributed to making this book possible.
I apologize in advance for any accidental omissions, you know who you are
and how much you are appreciated

Kent Knowles at the Raptor Conservancy of Virginia
Barbara Prescott, Erika Yery, and Dawn Davis at the Wildlife Rescue League
Peggy Bowers at the American Horticulture Society
Carolyn Gamble and Harry Glasgow at Huntley Meadows Park
Jen Schill, and Matthew Logan at the Potomac Conservancy
Mike Zimmerman at S&S Graphics
Roo Johnson, Andres Tremols, and John Bockelman at Vivo design
Ken Ferebee at Rock Creek National Park
Jessie Cohen at the National Zoo
Denise Ryan
Bill Folsom
Cristina Mittermeier
Michele McRae
Harry Chernoff
Dona Eichner

And my Dad for getting me started with a camera.

All of the photographs in the first two chapters, except for the shots of Great Falls, and two shots of turtles, were taken from my kayak on the Potomac river. Most of the photos in this book were taken with digital cameras. All the images were processed in Photoshop to crop them and make minor adjustments to color, contrast, etc... None of the photos were digitally manipulated or faked in any way...except one. Can you guess which? The answer is below.

Many of the images in this book are due to my being in the right place at the right time. Wildlife photography means spending a lot of time outside under the right lighting and weather conditions and waiting for something interesting to happen. With some luck interesting things do happen, and with even more luck I get the shot. But some of the photos were the result of having a vision for the kind of photograph I wanted, then trying to make it happen. This often involved bribing animals with a bit of food. The photos of the gulls in front of the Capitol and the Washington Monument happened this way. It also helps that gulls are not particularly shy around people.

However, one photo I had in mind proved impossible without taking some artistic license. For a long time I wanted to get a shot of a cardinal on top of the Vietnam Veteran's Memorial. The problem with this is that the memorial is very popular and never without a lot of people walking by it during daylight hours. No amount of enticement with food could get a shy cardinal to stop by. In the end I decided to combine a shot of the wall just after it snowed with a shot I had of a cardinal in the snow. The resulting photo is very close to what I had envisioned. If anyone, especially Vietnam veterans, feels that this is cheating, I apologize.

INTRODUCTION

Most people think the wild side of Washington can be found on Capitol Hill or in the bars of Georgetown and Adams Morgan, but the real wildlife in Washington is right in our own backyard. Millions of tourists come to Washington every year to see the Smithsonian, the cherry blossoms, the monuments, and the National Zoo, and rightly so, but what they don't see are the countless examples of natural beauty that are all around this city. This book will give you a glimpse into the real wildlife that Washington truly has to offer, not the elephants and donkeys that represent the man-made wildlife at the heart of this city.

The key natural feature that defines the city of Washington is the Potomac river. The Potomac is considered the wildest urban river in America. Except for two small dams just north of Washington DC that help provide drinking water to the region, the Potomac is free flowing for more than 380 miles. It is this wildness that allows a great diversity of wildlife to make their home on or near the river. This book is my attempt to document some of this wildlife.

This book started to take shape one day while I was kayaking on the Potomac river just outside Washington DC. The Potomac is a marvelous place to see wildlife and every time I got on the river I saw something amazing. And naturally each time I saw an amazing sight I did not have a camera with me. I decided to start taking my old 35mm camera to capture what I saw. The early results were not promising. My search for local wildlife expanded beyond the river to several local hot spots, backyards—including my own, and the Chesapeake Bay. Several years and lots of expensive camera equipment later I had enough wildlife photos for this book.

Thousands of people drive along and commute over the Potomac every day yet never stop to look at the animals making their homes there. These same people go home after work and never stop to look at the animals in their own backyards. In the summer, people drive over the Chesapeake Bay heading for the beach, but never notice the wildlife along the way.

I wanted this book to show people just how much natural beauty is around them every day, if they simply took the time to notice, and to encourage them to protect the natural habitat that gives wildlife a chance to survive.

Don Chernoff
September 2005

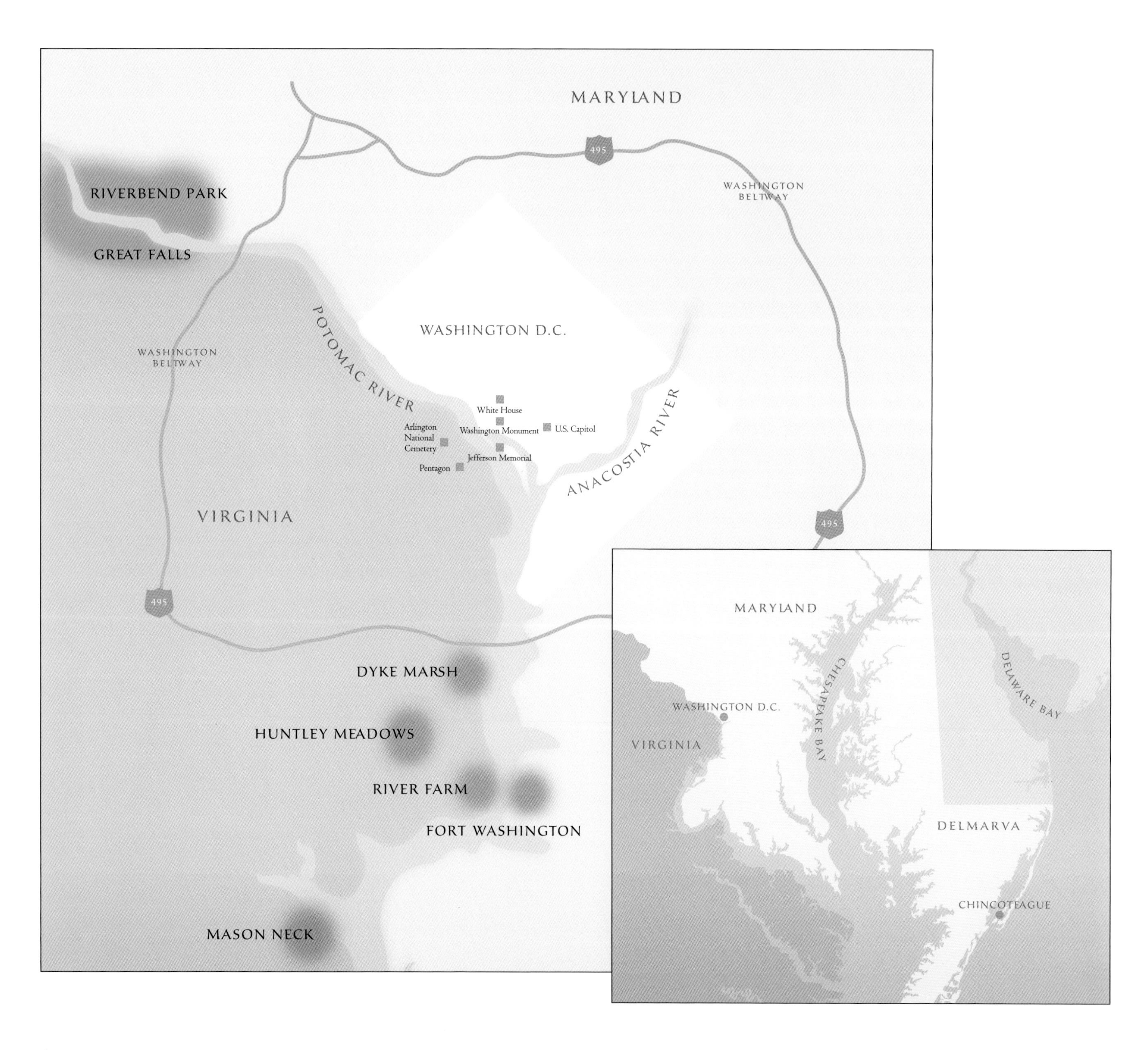
MARYLAND
495
WASHINGTON BELTWAY
RIVERBEND PARK
GREAT FALLS
POTOMAC RIVER
WASHINGTON D.C.
WASHINGTON BELTWAY
White House
Arlington National Cemetery
Washington Monument
U.S. Capitol
Jefferson Memorial
Pentagon
ANACOSTIA RIVER
VIRGINIA
495
495
DYKE MARSH
HUNTLEY MEADOWS
RIVER FARM
FORT WASHINGTON
MASON NECK
MARYLAND
CHESAPEAKE BAY
DELAWARE BAY
WASHINGTON D.C.
VIRGINIA
DELMARVA
CHINCOTEAGUE

TABLE OF CONTENTS

UPPER POTOMAC

RIVERBEND PARK TO GREAT FALLS

Great Falls on the Potomac, viewed from the Virginia side of the river. The falls create the closest world-class rapids to any major city in the United States. For this reason, the U.S. Olympic kayak team can often be seen training at the base of the falls. Riverbend Park lies just upstream from Great Falls.

A trio of male Wood Ducks hanging out on a log in the Potomac. Wood Ducks are the most vibrantly colored of all the ducks that visit the area.

A female Wood Duck taking off. Wood ducks are one of many species where the male and female look completely different, almost unrelated. This phenomena is known as dimorphism (two forms).

The Muskrat is a common mammal seen in or near the water. Sometimes mistaken for a Beaver, the Muskrat is smaller and lacks the large flat tail characteristic of the Beaver.

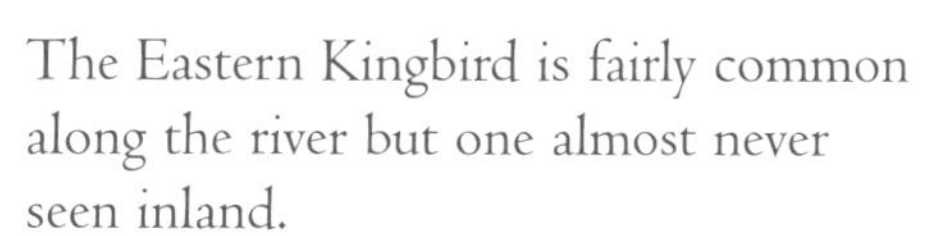

The Eastern Kingbird is fairly common along the river but one almost never seen inland.

The section of the Potomac within the boundaries of Riverbend Park, as seen from a kayak just before sundown. At the far end of this photo is the Washington Aqueduct. This damn turns the section of the river into something more like a lake than a wild river. The result is a relatively calm stretch of river that is ideal for wildlife that likes calmer water. The flat water also makes taking photos from a kayak much easier. The rapids of Great Falls are about a mile downstream from this spot.

Damselflies performing their mating ceremony on the water in late fall (right). Large clumps of damselflies can bee seen clinging to bits of wood or leaves as they float downstream

It is easy to see how the Zebra Swallowtail butterfly gets its name. With its unique black and white stripes, it is one of the most distinct butterflies in the region. This butterfly is puddling to collect nutrients, mostly salt, a behavior most common in males.

The Snowy Egret isn't seen as often as its larger cousins, the Great Egret and Great Blue Heron.

A Green Heron preening along the river's edge. Like all Herons and Egrets, Green Herons eat mostly fish. They can be seen wading in shallow water looking for a meal.

The Green Heron is unusual among Herons and Egrets as it typically isn't bothered by people getting close.

These pictures capture the moment of impact as the Green Heron extends its neck to grab a fish. All Herons and Egrets have extremely long necks that are normally folded up in an "S" shape. They extend their neck and almost double their body length when striking at food.

Dragonflies are very commonly seen near water. They often perch just long enough to get pictures like this. Photographing them in flight is practically impossible. This Blue Dasher is one of more than 100 species in the area.

It is more typical to see Tree Swallows making amazing acrobatic maneuvers over water as they catch insects in flight than perching as it is in this shot. If photographing a dragonfly in flight is difficult, photographing a Tree Swallow in flight is downright impossible.

A Red Spotted Purple is one of many colorful species of butterfly that can be easily be found if its proper food source is nearby.

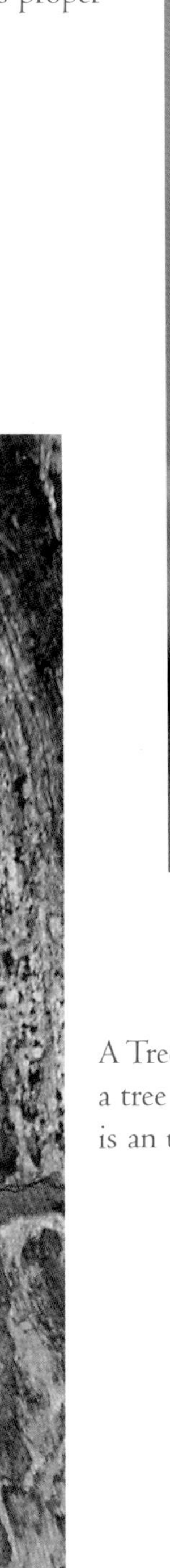

A Tree Swallow peering out of a cavity of a tree branch hanging over the Potomac river is an unusual sight.

Two views of Great Falls from approximately the same spot on the Virginia side.

What a difference a year makes. This view shows the falls under drought conditions in the spring of 2002.

This view shows the falls in September of 2003 a few days after hurricane Isabelle hit the region. A day earlier, the water was at least 10 feet higher. You can see the higher waterline as a dark line on the rocks along the right side of the photo.

A pair of Black Vultures take a break on a rock in the middle of the river. They are the smaller cousin to the more common Turkey Vulture. Black Vultures have an all black head which makes them easy to distinguish from the red or bright pink head of the Turkey Vulture.

A Turkey Vulture in a rarely seen position, not flying. These birds spend most of their day soaring high on thermals, looking for food. If you see one on the ground it is usually feeding. You can see them perched like this early in the morning or late in the day if the winds are calm and the air is too cool to generate the thermal currents they need to soar.

A very tiny American Toad, no bigger than a thumbnail, perfectly camouflaged in the leaves.

A Spotted Sandpiper with an insect snack. This is one of many species of birds that migrates through the area in the spring and can only be seen for a few weeks before it heads further north to nest.

LOWER POTOMAC

DYKE MARSH TO MASON NECK

An Osprey nests on a buoy used to construct the new Woodrow Wilson Bridge, part of the infamous Beltway around Washington, DC.

A pair of Herring Gulls relaxing on a rock in the Potomac next to the Woodrow Wilson bridge. The birds seemed unconcerned about the thousands of cars whizzing by on the beltway over their heads. The Washington Monument is in the background.

A pair of Double Crested Cormorants on a tree at Dyke Marsh. There were dozens of Cormorants in this tree, a sure sign that the river has plenty of fish for them to eat.

A Least Bittern hiding in Dyke Marsh, is a shy and rarely seen bird. When approached they freeze in place, hoping their camouflage makes them invisible.

A young Northern Water Snake basking on a log. These snakes are excellent swimmers and can often be found in the Potomac. The shiny appearance of this snake suggests that it recently shed its skin.

An Osprey bringing in a stick to finish building the nest while her mate can be seen poking his head up from the nest. Dyke Marsh is home to at least half a dozen breeding pairs of Osprey.

It is very common to see an Osprey with a fish, but rarely the whole fish. If you see one like this then you know the fish is freshly caught.

With 4 razor sharp talons on each foot, the Osprey is perfectly adapted for catching fish. This is a typical view of an Osprey flying with a headless fish. Osprey always eat the head of the fish first, maybe they know more about nutrition than we do.

A Beaver paddling across Dyke Marsh. Beavers are a very common sight in wetlands and calm areas of the Potomac.

A Black Crowned Night Heron on a sailboat in Belle Haven Marina next to Dyke Marsh.

A Spicebush Swallowtail butterfly feeding on a Button Bush.

A female Common Merganser resting on a log in Dyke Marsh (above) while her mate swims nearby (right).

Great Blue Herons are very common near water, but don't normally tolerate humans approaching them too closely. Sometimes animals like this tolerate a person in a boat or kayak more than they would a full human figure.

A Peregrine Falcon high in a tree overlooking Dyke Marsh. Peregrines are the fastest birds in the world and can dive at over 200 miles per hour to kill other birds in flight.

American Coots attempting a sunrise takeoff at Dyke Marsh. Coots are just about the least aerodynamic birds around so it takes them quite a while to get airborne.

A Great Blue Heron flies off into a muted sunrise over Dyke Marsh.

An immature Common Tern at Fort Washington marina on the Maryland side of the Potomac, just across the river from Mount Vernon.

A Gray Fox near Fort Washington. The area has two types of fox, Gray and Red. The Red Fox is the larger of the two.

Just downriver from Dyke Marsh on the Virginia side is Mason Neck State Park. Here a Painted Turtle suns on a log. Several species of turtle are very common on calm waters and can usually be found in a similar position enjoying the sunshine.

You might think that with all their armor, turtles are fearless. The truth is that they are very wary of humans and will usually drop into the water to hide before you ever see them. Often the only way you know turtles are near is when you hear the loud "plops" as they splash into the water.

An immature Bald Eagle on a tree over Kane's Creek in Mason Neck State Park. Bald Eagles do not develop the characteristic white head and tail feathers until about four years of age. Bald Eagles are extremely wary of humans and normally fly off at the first sign of human activity. This eagle was more cooperative than most.

A Belted Kingfisher perched over the water at sunset, another species that flees at the first sign of humans. They have a very noisy call that sounds something like a machine gun. Normally you hear them but only catch a glimpse as they are flying away.

An Eastern Box turtle rescued from the road outside Mason Neck. Box Turtles inhabit dense woodlands, unfortunately their habitat is disappearing at an alarming rate.

Sunset over Belmont Bay seen from Kane's Creek in Mason Neck State Park.

HUNTLEY MEADOWS

Huntley Meadows is a 1,400-acre refuge of wetland and woodland in Alexandria, Virginia. Because of its unique habitat and the easily accessible boardwalk and trails that run through the park, it is the best spot in the Washington area to view wildlife.

Two views of the rarely seen American Bittern. This secretive wetland bird normally hides among tall grasses and reeds. When frightened, it stands absolutely still with its head pointed skyward and uses its coloration to blend in.

A Common Snipe looking for food during a late winter thaw.

A Barn Swallow taking a break from gathering mud for its nest. Barn Swallows build a cup-shaped nest out of dried mud.

A Grasshopper on a Thistle.

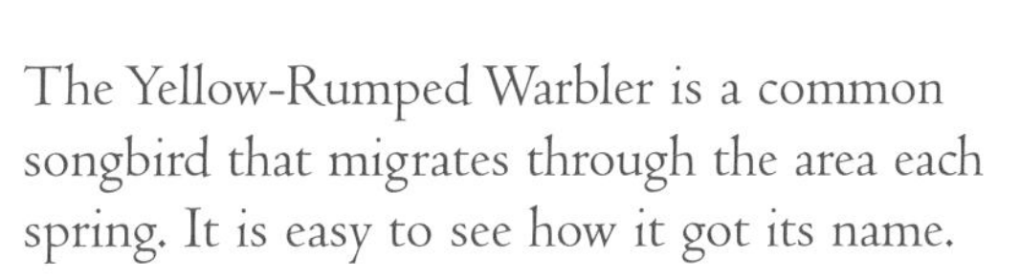

The Yellow-Rumped Warbler is a common songbird that migrates through the area each spring. It is easy to see how it got its name.

A Great Egret catching a mid-day snack. These large birds can catch fish so large that they choke trying to swallow it whole. This small fish did not present any problem.

A Bullfrog is one of the more common sights at Huntley Meadows.

A young painted turtle finds an interesting way to catch some rays. How he got up there I'll never know.

A Red-headed Woodpecker high in a snag. Snags are dead trees that provide food and nesting sites. They are critical to the survival of woodpeckers and many other animals.

A male Red-winged Blackbird singing in the wetland. Blackbirds have a distinct song that is almost the signature sound of a wetland.

A female Red-winged Blackbird is neither black nor has red feathers on her wings. Another example of dimorphism, an adaptation that allows the females to better blend in with their environment.

A Dark-Eyed Junco foraging in the snow. Juncos are ground feeders and can often be seen digging for seeds.

A Bullfrog in a less than desirable position, as lunch for a hungry Northern Water Snake. This frog was considerably larger than the snake's head and it took about 20 minutes for the snake to swallow such a large meal.

A Killdeer is an unusual shorebird with an unusual name. It is one of the only shorebirds that you are more likely to see inland than at the beach.

A Sora feeding in the wetland.

A Song Sparrow hiding out among the reeds.

A Northern Mockingbird feasting on the last berries of the season, just before the first winter snow.

A Canada Goose dad keeping the newly hatched chicks warm while the mom was feeding nearby. These chicks were only a day or two old. Female adult Geese are larger than males.

Canada Goose triplets from a different brood a few weeks later.

A very common resident of the wetland is the Snapping Turtle. Here is a rare sight of two snapping turtles mating. The male is on top and he is holding the female mostly underwater.

Another rare glimpse of a snapping turtle, this time walking on dry land. Snappers are creatures of the water and are more typically seen submerged with only their nose sticking out of the water to breathe.

Believe it or not, this is the same view as the opening photo of this chapter, about two years later, in the summer of 2005.

The Beavers that created this wetland decided to move downstream and build a new damn and lodge.

Beavers are called "nature's engineers". They create wetlands to provide food and shelter for themselves. When their food source is depleted, they move on and start a new wetland, giving the old land a chance to regenerate.

The new beaver lodge sits in the center of
the new downstream impoundment.

WASHINGTON DC

THE MALL, SMITHSONIAN, THE NATIONAL ZOO, AND ROCK CREEK PARK

A Ring-billed Gull poses in front of the Capitol building. The reflecting pond in front of the capitol is a popular spot for many water birds.

A pair of Mallard ducklings swim in the wetland at the Smithsonian's National Museum of the American Indian. The building was designed to look like it was carved out of stone cliffs. In keeping with the theme of a natural landscape, museum designers created this wetland at the east end of the building. It is a tiny oasis in the middle of the mall that is home to many animals not normally seen in an urban area.

In the wetland in front of the Indian Museum a Red-winged Blackbird sings to defend its territory. In the background is the Statue of Freedom on top of the Capitol building.

A breeding pair of Mallard Ducks swims in a pond behind the Smithsonian's first building, popularly known as "the Castle". Completed in 1855, it now houses a visitor's center.

A flock of European Starlings resting on top of the statue at the top of the Smithsonian's Arts & Industries building. This building, opened in 1881, was the original home of what was then called the National Museum.

A female Northern Cardinal in the snow atop the Vietnam Veteran's Memorial.

Ducks in front of the Jefferson Memorial, with the famous cherry blossoms blooming.

A Ring-billed Gull and Rock Dove, commonly called a Pigeon, have a heated discussion, similar to what happens inside the Capitol building when Congress is in session, yet undoubtedly more civilized.

A Northern Rough-winged Swallow (left) relaxes while a Double-Crested Cormorant (right) swims in the Tidal Basin in front of the Jefferson Memorial. The columns of the memorial are reflected in the water.

A male American Wigeon is a rare visitor to this pond, normally frequented by gulls, Mallard ducks, and Canada geese.

An early spring Robin gets the worm. This Robin is standing on the wall surrounding the Smithsonian Sculpture Garden. The statue in the background is "Lunar Bird" by Joan Miro.

A Female Mallard Duck takes a break at the Smithsonian Sculpture Garden underneath a mobile by Alexander Calder. The Sculpture Garden is part of the Hirshorn Museum of Modern Art. This mom had twelve ducklings so she really needed this rest.

A Ring-billed Gull swoops in for a well-timed piece of bread
over the pond in front of the Washington Monument.

A juvenile Black-crowned Night Heron, exhibiting unusual behavior, hangs upside down in a tree next to the bird house at the Smithsonian Institution's National Zoo.

An interesting fact about the National Zoo is that a lot of wild animals make their home there. Here is a famous Black-crowned Night Heron rookery which, interestingly enough, is located right next to the bird house.

An American Crow hanging out at the National Zoo.

A black-phase Gray Squirrel enjoys the abundance of food to be found on the grounds of the National Zoo.

The abundance of wild prey like squirrels and birds probably attracted this Red-shouldered Hawk to the Zoo. Or maybe he just liked hanging around the other birds

A male Pileated Woodpecker with a beak full of wood from excavating a nest cavity in Rock Creek Park. At the time this photo was taken this was the largest woodpecker in north America. A few weeks later its larger cousin, the Ivory-Billed Woodpecker was rediscovered in Arkansas after having been thought extinct for 60 years.

Without dead trees to nest in, the Pileated Woodpecker would suffer the same fate.

BACKYARDS

A male American Goldfinch. In the spring Goldfinches, like many birds develop breeding plumage that is more colorful than during most of the year. Male Goldfinches are almost fluorescent yellow, like a small tennis ball with wings. Females are a much duller yellow.

The very common Mourning Dove. This dove gets its name from the soulful sound of its call.

A White-Breasted Nuthatch in a typical position, upside down on a tree. Nuthatches cling to trees like woodpeckers, but are not related to them. Only the Nuthatch has the unusual habit of facing head-first down a tree.

A female Red-Bellied Woodpecker checking out the landscape. You can see the patch of red feathers on the abdomen that gives this species its name.

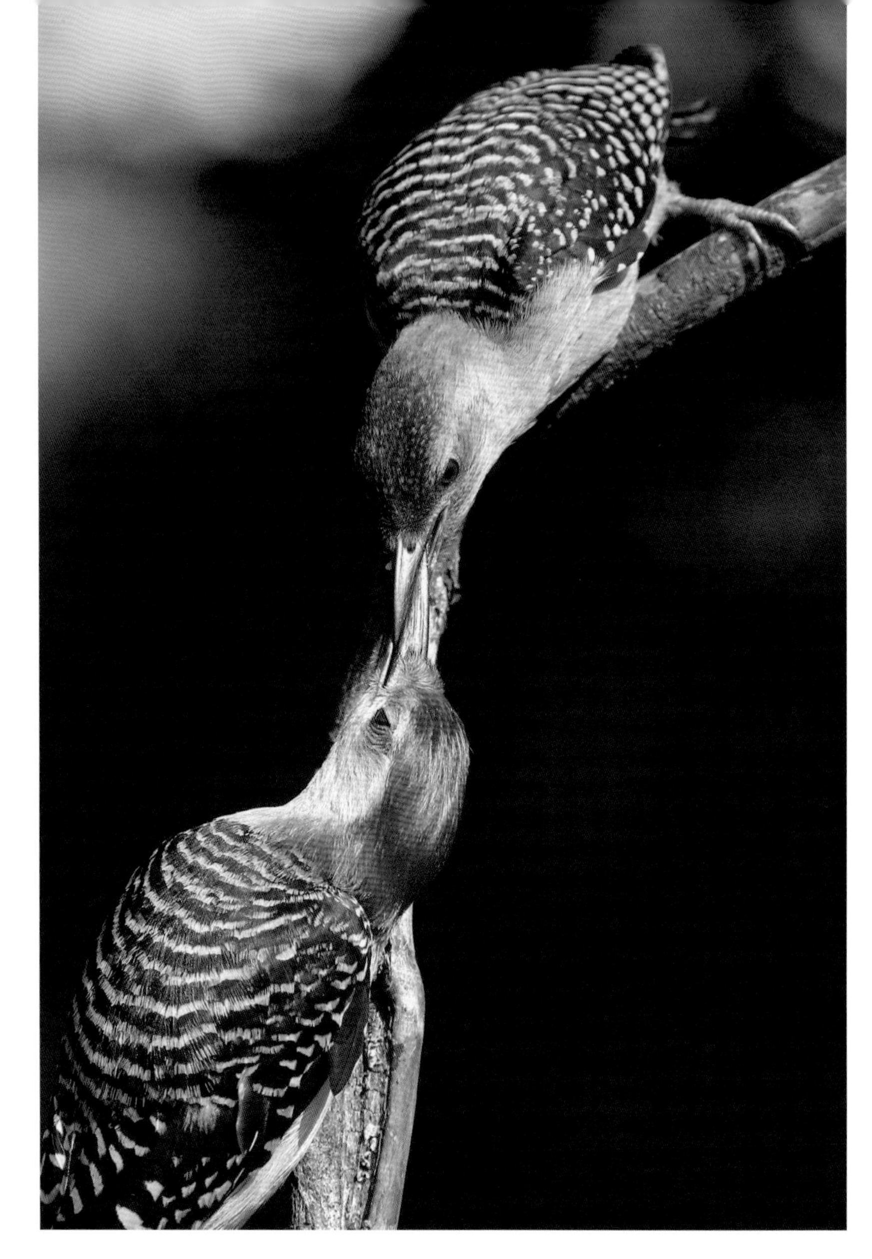

A male Red-Bellied Woodpecker (bottom) feeding its offspring. Males have red feathers extending completely over their head while females have only a partially red head.

A Carolina Chickadee feeding its young. This tiny bird might be, ounce for ounce, the most fearless bird around. They don't appear to be afraid of people and will come to feeders only a few feet away from humans.

This Carolina Chickadee fledgling was calling for its parents to bring food. The yellow coloration on the bill is typical of many newly hatched birds and fades to the normal color as the bird matures.

Cicadas were big news in the summer of 2004. There are many species of Cicada but these were from "Brood X" and they emerge only once every 17 years. They stick around for several weeks, make ridiculous amounts of noise, and smell terrible when they all die en masse after breeding.

On the left is an adult a few days after emerging from it nymph phase and leaving behind its empty exoskeleton (right).

One of the great things about Cicadas, from a photographer's perspective, is the ability to pick them up and pose them in just about any position you desire.

A male Northern Cardinal stares down the camera. The Northern Cardinal is the state bird of Virginia.

A Tiger Swallowtail feeding on Waxleaf Privet. The Tiger Swallowtails is the state insect of Virginia.

The dark form of the female Tiger Swallowtail

An American Painted Lady butterfly on Waxleaf Privet.

Carolina Wrens are famous for making their nests in the strangest places. This pair picked an empty deli container in my carport as their dream home site, next to an old jug of anti-freeze. Here one of the adults is bringing a tasty insect to feed the chicks.

The Gray Catbird makes a call that sounds very much like the meow of a house cat.

A Northern Mockingbird with a spider snack. Mockingbirds imitate the calls of the birds around them. It is thought that they can mimic hundreds of other birds' calls.

A male Downy Woodpecker doing what woodpeckers do. The Downy Woodpecker is the smallest woodpecker in North America. Only the male has a red patch of feathers on the back of its head.

A male Baltimore Oriole feeding its chick in the nest. Orioles make a nest that is basket of plant fibers that hangs from the end of a tree branch.

The Baltimore Oriole is the state bird of Maryland.

The same Baltimore Oriole chick two days later, stretching its new feathers after emerging from the nest.

A Tufted Titmouse in the snow.

A HELPING HAND

WILDLIFE REHABILITATORS

A pair of month-old Raccoon siblings being raised at the Wildlife Rescue League (WRL)
All the mammals shown on the next three pages are courtesy of the WRL.

A baby Chipmunk, only a few weeks old, hasn't yet opened its eyes.

Hard working volunteers at the Wildlife Rescue League care for many kinds of orphaned animals and raise them to adulthood. Once the animals can survive on their own they are returned to the wild.

The same Chipmunk a little over a month old. This guy will be released into the wild at about two months of age.

A very uncommon view of a pair of very common Gray Squirrels. Squirrels raise their young in a nest, much like birds. The babies stay in the nest until they are almost fully grown.

An older Gray Squirrel, but still younger that what would normally be seen in the wild. This "teenager" was afraid of venturing much beyond this perch. A fully grown squirrel ready for release would probably have run away.

A Southern Flying Squirrel is another common local animal that most people never encounter because the squirrels are nocturnal. Scientists estimate that the population of Flying Squirrels is even larger than the population of Gray Squirrels.

Flying Squirrels have large flaps of skin on each side that can be extended to make a sort of parachute that lets them jump from high in a tree and glide gently to nearby trees.

The Barred Owl is commonly heard but rarely seen. The name comes from the alternating brown and white bars on its chest and wings.

This owl has an injury to its wing and is a permanent resident at the Horsehead Wetland's Center.

The Horsehead Wetlands Center on Maryland's Eastern Shore is home to this pair of injured Bald Eagles. Because of injuries to their wings, these eagles are not releasable and are used to educate the public.

The Eastern Screech Owl is common but rarely seen because it is nocturnal and its small size and coloration lets it blend in well with trees.

Owls, hawks, and falcons are raptors. The following raptors were injured and cannot be returned to the wild. They are all cared for at the Raptor Conservancy of Virginia (RCV) and used in education programs at schools.

Screech Owls exhibit a great variation in their plumage. This is a red-phase Eastern Screech Owl. At about eight inches tall, Screech Owls are the second-smallest owls in the DC-area.

A Barn Owl lets out a terrifying scream. Owls hunt at night and have special feathers that let them fly without making noise.

This silent flight and the generally white coloration of the Barn Owl, along with its blood curdling scream, are thought to be the origins of the myth of ghosts flying through the dark night.

A Peregrine Falcon enjoying a meal.

Peregrines naturally nest in high rocky cliffs and catch birds in flight. They can dive on and kill prey at over 200 miles per hour.

These falcons have been seen nesting in cities along the east coast, where tall buildings simulate rocky cliffs and the abundance of pigeons makes for an easy meal.

The American Kestrel is the smallest falcon in North America.

The Short-eared Owl can sometimes be seen hunting during the day.

A Red-tailed Hawk is the largest and most common hawk in the region. In flight, the underside of the tail feathers appear white with dark bands, similar to other hawks. This bird is used for education. The straps on its ankles, called jesses, are a type of leash used to control the bird.

The Great-horned Owl is the largest owl and possibly the most feared predator in the region. It is also the most widespread owl in North America. Great-horned Owls are powerful hunters and will attack prey larger than themselves. They will also kill and eat other owls.

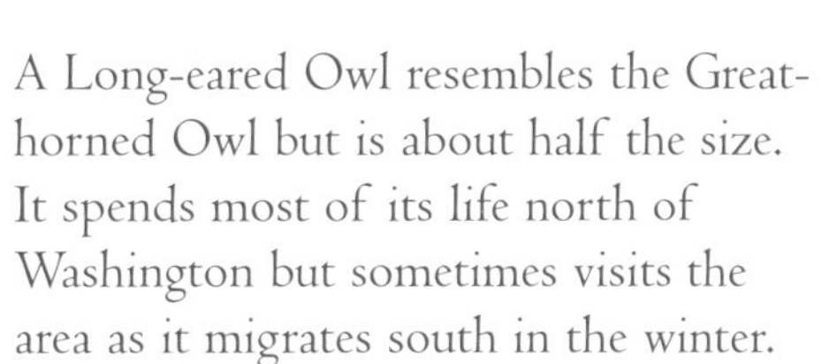

A Long-eared Owl resembles the Great-horned Owl but is about half the size. It spends most of its life north of Washington but sometimes visits the area as it migrates south in the winter.

The Northern Saw-whet Owl is the smallest owl in the area but it has the same attitude and instincts of its larger relatives. It eats small rodents and large insects.

An up-close look into the eye of a Great-horned Owl. An owl's eyes are more than 100 times as sensitive to light as our own.

RIVER FARM

AMERICAN HORTICULTURAL SOCIETY

River Farm is a 27-acre estate once owned by George Washington, just up the road from his more famous home: Mount Vernon. Since 1973 it has been home to the American Horticultural Society. Thanks to its location on the Potomac River, its unique mix of habitats, and its diverse flower gardens, River Farm is a kind of "ultimate backyard" that attracts a wide variety of wildlife.

A young Red Fox kit exploring the world beyond the family den. The Red Fox is larger and more common than its relative, the Gray Fox.

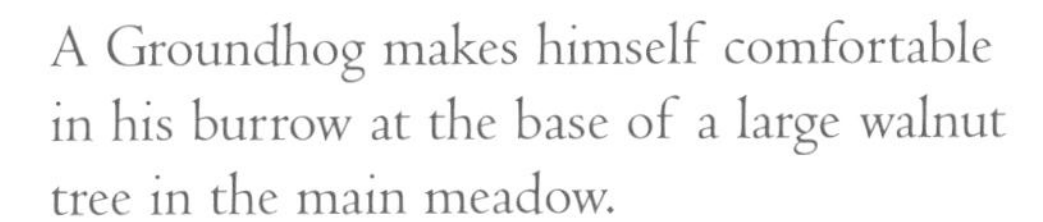

A Groundhog makes himself comfortable in his burrow at the base of a large walnut tree in the main meadow.

A Monarch Butterfly feeding on the aptly named Butterfly bush.

Monarchs also eat the toxic nectar of the Milkweed plant which makes them distasteful to birds and other predators.

The Palamedes Swallowtail is one of the largest butterflies in the area. This one is feeding on a Zinnia. Swallowtails are a family of large colorful butterflies named for the tail-like projections on their hind wings.

A male Eastern Bluebird pauses at its nest box before delivering a tasty fly to its hatchlings. The males are much more brightly colored than the females.

On this day the female won the prize for bringing the largest main dish for dinner. The screen surrounding the opening to the next box prevents larger birds from entering.

A Painted Lady butterfly enjoying some Lantana nectar. The Painted lady is a very close relative of the American Painted Lady on page 68.

A male Brown-headed Cowbird overlooking the main meadow. Cowbirds exhibit a behavior called nest parasitism. They lay an egg in the nests of other birds and expect these birds to raise the Cowbird chick. Because this behavior is helped by land clearing and fragmentation, suburban development has caused an increase in the Cowbird population at the expense of other birds.

A Brown Thrasher takes cover in a tight spot.

A female Ruby-throated Hummingbird gets a last meal of Macleaya Poppy nectar before migrating south in the fall. A week later this plant had withered from the cold.

A Pink-Edged Sulphur Butterfly feeding on a New York Aster. Sulphur is the family name for a large variety of small yellow butterflies.

A Great Spangled Fritillary Butterfly enjoying a meal of Echinacea, also called Purple Coneflower.

The flower gardens at River Farm are used to educate people about just how easy it is to attract a wide variety of beautiful butterflies to your yard if you have the plants they like to feed on.

The Wild Turkey was a familiar site when George Washington was the owner of this land. Hunting and habitat destruction have made this a rare bird in the wild. It is a testament to the habitat preservation at River Farm that Wild Turkeys still roam the grounds. This young male is over three feet tall.

A Pearl Crescent butterfly feeding on a Black-eyed Susan.

DELMARVA PENINSULA

CHESAPEAKE BAY AND THE EASTERN SHORE

Sunset over Chincoteague National Wildlife Refuge at the southern tip of the Delmarva peninsula. Delmarva is made up of pieces of Delaware, Maryland, and Virginia. The peninsula, known to Washingtonians as the Eastern Shore, separates the Chesapeake Bay on the west from the Delaware Bay and Atlantic Ocean to the east.

A White-tailed Deer buck at Sandy Point State Park, MD. The park is at the western edge of the Chesapeake Bay. The Chesapeake Bay Bridge to the Eastern shore is in the background.

The other half of the happy couple, a White-tailed deer doe, in front of the Chesapeake Bay Bridge.

A row of Ruddy Turnstones overlooking the Atlantic Ocean. Ruddy Turnstones get their name because of their habit of turning over rocks in search of food.

A young Great Black-backed Gull feeds on a Horseshoe Crab on a beach in Delaware.

For millions of years these prehistoric crabs, more closely related to spiders than true crabs, have laid their eggs on the beaches of the Delaware Bay. Many species of Gulls and shorebirds have evolved their migration patterns to arrive at these beaches in the spring to feed on the Horseshoe Crab eggs.

These shorebirds migrate from South America to nest in the Arctic. Without this important refueling stop these birds would not survive to make it to their breeding grounds.

Unfortunately for many migrating shorebirds, people have upset the balance of nature by over-harvesting Horseshoe Crabs to use as bait.

Without the super abundance of Horseshoe Crab eggs to feed on each spring, many migrating birds have suffered terrible losses.

The poster child for this problem is the Red Knot. Because of the decrease in their critical fuel supply, Red Knot populations have plummeted and this bird may soon be added to the endangered species list.

A Black-neck Stilt searches the shoreline for food. These elegant birds, like most shorebirds, are adapted for probing the sand and shallow waters to find food. Because the Stilt has longer legs than most shorebirds, it can forage for food in deeper water.

It is unusual to find a Fowler's toad on the beach. This toad's coloration allows it to blend in with the sand. Without this camouflage it would become an easy meal for the gulls.

A Least Sandpiper is one of many similar looking shorebirds that migrate to the Delaware Bay shoreline to feed on Horseshoe Crab eggs.

A pair of Laughing Gulls taking a break. These gulls are very numerous near the beaches and get their name from their loud call that sounds like laughter.

A Ruddy Turnstone takes to the air to escape a large wave.

A Sanderling keeping a close eye on the tide to see if it exposes something tasty in the sand.

A young male White-tailed Deer trying to hide in the tall grass opposite the beach along the Delaware shoreline. His antlers are just starting to grow and are covered by soft fur called velvet.

Though not a shorebird, this Savannah Sparrow tries to get in on the feeding action.

A Semipalmated Sandpiper finds a dry spot from which to survey the beach for food. This is another species that nests in the Arctic and must fatten up on Horseshoe Crab eggs to survive the trip.

A good place for this Willet to look for eggs is in front of a pileup of Horseshoe Crabs.

Most of the millions of crabs that come ashore return to the ocean, but that still leaves plenty that get stranded and die on the beach.

A serious-looking Great Egret in the Chincoteague National Wildlife Refuge.

A Marsh Wren singing in a thicket along the shoreline.

A Sika Deer at dawn in the Chincoteague National Wildlife Refuge. Sika Deer are a type of oriental elk that were introduced into the refuge in the 1920's. They are much smaller than the native White-tailed Deer.

ENVIRONMENTAL ISSUES

WHAT WE CAN DO TO HELP WILDLIFE

It only took 15 minutes to collect all this trash floating in the Potomac river near Georgetown. The entire watershed of the Washington DC area empties into the Potomac, and the smaller Anacostia River, which in turn empty into the Chesapeake Bay. In addition to its wonderful habitat for wildlife, the Potomac also supplies the drinking water for most of the 5 million people who live in the Washington metro area. We all need to do a better job of keeping our rivers clean.

This Double-crested Cormorant died a horrible death due to the negligence of a fisherman. The bird most likely was attracted to the bait on a fishing line that got snagged in a tree.

Fisherman benefit from a clean river and have an obligation to keep the river clean and safe for all wildlife.

What's wrong with this picture? This photo is of a mature forest along Scott's Run nature preserve bordering the Potomac. It looks like a healthy forest, but looks are deceiving. A normal forest this time of year would be full of undergrowth, small shrubs and saplings growing from seeds dropped from the mature trees. It is difficult to walk through a healthy forest.

This forest looks almost barren of plant life below approximately six feet high. The reason for this is an overabundance of White-tailed Deer. Deer eat almost all vegetation up to about six feet in height. When there are too many deer for a forest to support, the result is a forest with no undergrowth. This may not seem like a big problem now, but a few decades down the road when the mature trees reach the end of their life there will be no trees to replace them, in other words this forest is dying. A bigger problem in the short term is that dense undergrowth provides habitat and food for dozens of other animals. If deer are allowed to strip a forest clean, species that rely on the forest for survival will disappear.

The solution, as harsh is it may sound, is to reduce the population of White-tailed deer. Support your local efforts to reduce deer populations to manageable levels. Those who resist efforts to control the deer population are only dooming our forests and the animals that depend on them to death. In the end, if the forests die so will the deer. For more information visit www.suburbanrenewal.org.

What's wrong with this picture? How can a beautiful green lawn be an environmental problem?

The number one problem affecting wildlife is habitat destruction from human development. Our suburban landscapes with their lush green lawns give the illusion of an oasis. In reality, they are nothing more than a biological desert that provide no food and no shelter for animals. In fact, because of all the pesticides, fertilizers, and other chemicals that are used to keep lawns looking nice, it would be better for the environment if they were replaced with green concrete.

What can you do? Replace your lawn with native plants that provide food and shelter for animals. Create a butterfly garden, it will require much less cost and maintenance than a lawn and provide you with the beauty and enjoyment of the butterflies it attracts. Dig up your lawn and let nature reforest it with seeds left by birds, squirrels and other animals. You will save time and money, but more importantly, you will save wildlife.

To learn more about how you can transform your lawn into habitat for wildlife, visit www.suburbanrenewal.org.

If you would like to learn more about the locations mentioned in this book, or would like to donate to organizations that help injured wildlife and help protect the local environment, here is a list of web sites where you can get more information

Raptor Conservancy of Virginia
www.raptorsva.org

Wildlife Rescue League
www.wildliferescueleague.org
www.wildbunchrehab.org

Huntley Meadows Park
www.fairfaxcounty.gov/parks/Huntley

The Potomac Conservancy
www.potomac.org

Great Falls National Park
www.nps.gov/grfa

Friends of Dyke Marsh
www.fodm.org

Mason Neck State Park
www.dcr.state.va.us/parks/masonneck.htm

National Audubon Society
www.audubon.org

The Nature Conservancy
www.nature.org

Chesapeake Bay Foundation
www.cbf.org

Suburban Renewal
www.suburbanrenewal.org

Book Layout by
VIVO Design
www.vivodesign.com